Sound Advice on

Equalizers, Reverbs & Delays

by Bill Gibson

236 Georgia Street, Suite 100
Vallejo, CA 94590
(707) 554-1935

©2002 Bill Gibson

Publisher: Mike Lawson
Art Director: Stephen Ramirez; Editor: Patrick Runkle

Cover image courtesy Midas.

ProAudio Press is an imprint of artistpro.com, LLC
236 Georgia Street, Suite 100
Vallejo, CA 94590
(707) 554-1935

Also from ProMusic Press
Music Copyright for the New Millennium
The Mellotron Book
Electronic Music Pioneers

Also from EMBooks
The Independent Working Musician
Making the Ultimate Demo, 2nd Ed.
Remix: The Electronic Music Explosion
Making Music with Your Computer, 2nd Ed.
Anatomy of a Home Studio
The EM Guide to the Roland VS-880

Also from MixBooks
The AudioPro Home Recording Course, Volumes I, II, and III
The Art of Mixing: A Visual Guide to Recording, Engineering, and Production
The Mixing Engineer's Handbook
The Mastering Engineer's Handbook
Music Publishing: The Real Road to Music Business Success, Rev. and Exp. 5th Ed.
How to Run a Recording Session
The Professional Musician's Internet Guide
The Songwriters Guide to Collaboration, Rev. and Exp. 2nd Ed.
Critical Listening and Auditory Perception
Modular Digital Multitracks: The Power User's Guide, Rev. Ed.
Professional Microphone Techniques
Sound for Picture, 2nd Ed.
Music Producers, 2nd Ed.
Live Sound Reinforcement
Professional Sound Reinforcement Techniques
Creative Music Production: Joe Meek's Bold Techniques

Printed in Auburn Hills, MI
ISBN 1-931140-25-1

Contents

Introduction

A musical work of art is a very complex combination of pieces integrated together in a way that conveys great emotion, passion and personality. Creating sonic ingredients that fit tightly together as a powerful and cohesive—yet musical—unit isn't always an easy task.

An excellent engineer strives at each step to maximize the integrity of the audio signal. An excellent producer strives at each step to maximize the integrity of the musical work. An excellent musician strives on each take to provide the very best and most musical performance possible. An excellent songwriter strives to craft each lyric into the most powerful and impacting word picture. Music is teamwork!

Each of us in the audio world must at some point grasp the concept that details matter. An inexperienced recordist/producer is often pleased to just get an audio signal recorded. Many tracks are printed as "close enough" or left until mixdown, where an attempt will be made to polish them up. At the same time, that recordist/producer wants their music to sound as good as their favorite recording. This approach is flawed.

If we want our music to compete with what we hear on the radio our standards must stay high at each stage. Always hold out for the best take, or the perfect sound, or the most passionate and emotional delivery. Remember, we're all competing with teams of wonderfully creative and talented people trying their hardest to take the art of music to the next level. I love a phrase that I picked up a long time ago from Andy Byrd at Warner Chappell in Nashville. Doing a fair amount of work together, we'd work through the parts with the players and singers until we had a great take, then he'd often come back with, "Okay, that's great! We'll save that one. Now let's take it a little further." I've adopted that philosophy in all I do. Get whatever you're doing to where it seems as good as you can get it. Then, reevaluate to see if there's any possible thing you can do to "take it a little further."

When considering equalization, reverbs, and delays strive to shape each sound into a finely honed puzzle piece. Sculpt away ingredients that aren't necessary and smooth off rough edges so each piece fits together to form a work that has strength, power, and integrity. Too often, we want to just turn a track up, then "throw a little reverb" on it or just "boost the lows a little" or perform other very general nonspecific operations. Then we wonder why the mix is muddy or why it sounds quiet while the levels look hot or why it's too boomy in the car, and on, and on.

This book contains techniques and information that will help you begin to finely craft your music for optimum power and a competitive edge in the marketplace. Pay attention to these details and you will soon see your music take a quantum leap forward in every way. That's cool! Now let's "take it a little further."

Integrating EQs, Reverbs and Delays

It'll be difficult to effectively utilize any of these tools if they're not connected and integrated in the manner that's most efficient for the application. Thorough understanding of the following types of component integration and signal path routing will give you confidence in the integrity of your audio signal.

Channel Insert

Most modern mixers have what is called a channel insert. This is the point where a piece of outboard signal processing can be plugged into the signal path on each individual channel. If your mixer has inserts, they're probably directly above or below the microphone inputs.

Channel Insert

Many mixers have a channel insert. This is the point where an outboard signal processor can be plugged into the signal path. A channel insert has a send that routes the signal, usually as it comes out of the preamp, to the processor. The output of the signal processor is then patched into the return of the channel insert. This completes the signal path. The signal then continues on its way through the EQ circuit and on through the rest of its path.

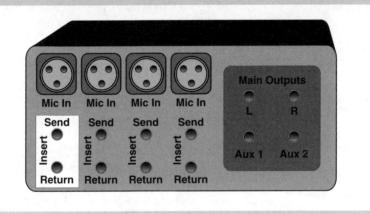

A channel insert lets you access only one channel at a time and is used to include a signal processor in the signal path of that specific channel. The processor you insert becomes a permanent part of the signal path from that point on. An insert is especially useful when using a compressor, gate or other dynamic processor.

Simple Send and Return Channel Inserts

With this sort of setup, the send and return are said to be normalled, because they are normally connected together.

Normalled Connections

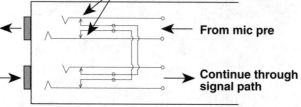

Contact at these points connects the sends to the returns when there's no jack inserted. The audio signal flows through this patch point unaffected until you plug into the jack.

From mic pre

Continue through signal path

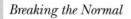
Breaking the Normal

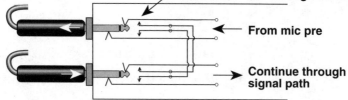

Plugging into the jack breaks the contact here. When these points don't touch, the send and the return aren't connected through the normal.

From mic pre

Continue through signal path

Jumpered Send and Returns

The send and return are only connected when the jumper is plugged into both jacks at once. If this jumper is removed, you won't hear the signal. When outboard processing is needed, simply remove the jumper, patch the send to the processor input, then patch the processor output to the return.

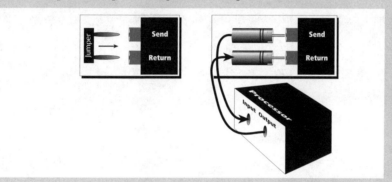

A channel insert utilizes a send to send the signal (usually as it comes out of the preamp) to the signal processor. The signal processor output is then patched into the return of the channel insert. This completes the signal path, and the signal typically continues on its way through the EQ circuit and on through the rest of its path.

The Single Insert Jack

To utilize this type of insert, you must use a special Y cable, like the one below (male tip-ring-sleeve stereo 1/4" phone plug to two female tip-sleeve mono 1/4" phone jacks). You can also use a special cable with one tip-ring-sleeve male and two tip-sleeve male connectors. Plug the male stereo phone plug into the insert. Next, use a line cable to connect one of the female mono connectors to the processor input, and patch the output of the processor into the other female mono connector. You might need to experiment to determine which of the mono connectors is the send and which is the return.

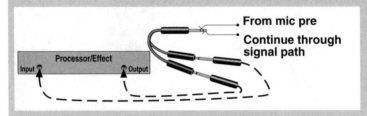

From mic pre

Continue through signal path

Processor/Effect
Input   Output

A channel insert and an effects bus are similar in that they deal with signal processing. An insert affects one channel only. Inserts are ideal for patching dynamics processors into a signal path. An effects bus (like aux 1 or aux 2) lets you send a mix from the bus to an effect, leaving the master mix on the input faders without effects. The output of the effect is then plugged into the effects returns or open channels on the mixer. This is good for reverbs and multi-effects processors.

The Effects Bus

When we discuss the input faders as a group, we're talking about a bus. The term bus is confusing to many, but the basic concept of a bus is simple—and very important to understand. A bus usually refers to a row of faders or knobs.

If you think about a city bus, you know that it has a point of origin (one bus depot) and a destination (another depot), and you know that it picks up passengers and delivers them to their destination. That's exactly what a bus on a mixer does. For example, in mixdown the faders bus has a point of origin (the tape tracks) and a destination (the mixdown recorder). Its passengers are the different tracks from the multitrack.

Aux Bus with Pre and Post

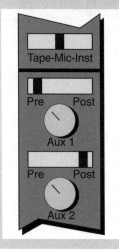

On Aux 1, the Pre-Post switch is set to Pre, letting the Aux 1 bus hear the signal from the channel input before it gets to the EQ and fader. On Aux 2, the Pre-Post switch is set to Post, letting the Aux 2 bus hear the signal from the channel input after it has gone through the EQ and fader circuitry.

Most mixers also have auxiliary buses, or effects buses. Aux buses (also called cue sends, effects sends or monitor sends) operate in the same way as the faders bus. An aux bus (another complete set of knobs or faders) might have its point of origin at the multitrack or the mic/line inputs. It picks up its own set of the available passengers (tracks) and takes them to their own destination (usually an effects unit or the headphones).

When a bus is used with an effect, like a reverb, delay or multi-effects processor, the individual controls on the bus are called effects sends because they're sending different instruments or tracks to the effects unit on this bus. The entire bus is also called a send.

Return is a term that goes with send. The send sends the instrument to the reverb or effect. The return accepts the output of the reverb or effect as it returns to the mix.

Pre and Post Fader

Aux buses often include a switch that chooses whether each individual point in the bus hears the signal before it gets to the EQ and fader (indicated by the word pre or pre fader) or after the EQ and fader (indicated by the word post or post fader).

Pre lets you set up a mix that's totally separate from the input faders and EQ.

This is good for headphone sends. Once the headphone mix is good for the musicians, it's best to leave it set. You don't want changes you make for your listening purposes to change the musicians' mix in the phones.

Post is good for effects sends. A bus used for reverb sends works best when the send to the reverb decreases as the instrument or voice fader is turned down. Post sends are perfect for this application since the send is after and dependent on the fader. As the fader is increased and decreased, so is the send to the reverb, maintaining a constant balance between the dry and affected sounds. If a pre send is used for reverb, the channel fader can be off, but the send to the reverb is still on. When your channel fader is down, the reverb return can still be heard, loud and clear.

Using the Aux Bus

Imagine there's guitar on track 4, and it's turned up in the mix. We hear the guitar clean and dry. Dry means the sound is heard without effects. The guitar in Audio Example 1 is dry.

Audio Example 1
Dry Guitar

If the output of aux bus 1 is patched into a reverb, and the aux 1 send is turned up at channel 4, we should see a reading at the input meter of the reverb when the tape is rolling and the track is playing. This indicates that we have a successful send to the reverb.

The reverb can't be heard until we patch the output of the reverb into either an available, unused channel of the mixer or into a dedicated effects return. If your mixer has specific effects returns, it's often helpful to think of these returns as simply one or more extra channels on your mixer.

Once the effects outputs are patched into the returns, raise the return levels on the mixer to hear the reverb coming into the mix. Find the adjustment on your reverb that says wet/dry. The signal coming from the reverb should be 100 percent wet. That means it's putting out only reverberated sound and none of the dry sound. Maintain separate control of the dry track. Get the reverberated sound only from the completely wet returns. With separate wet and dry control, you can blend the sounds during mixdown to produce just the right sonic blend. Listen to Audio Examples 2, 3 and 4 to hear the dry and wet sounds being blended in the mix.

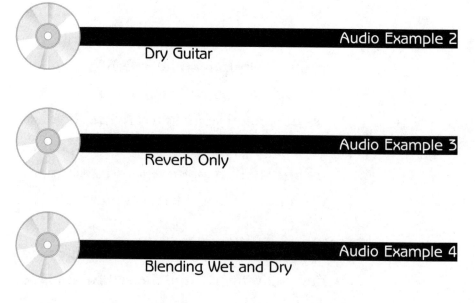

Audio Example 2

Dry Guitar

Audio Example 3

Reverb Only

Audio Example 4

Blending Wet and Dry

The Equalizer

Equalization principles apply to all equalizers. Parameter control options vary between EQ types but a thorough understanding of the principles behind EQ adjustments gives you the ability to instantly shape audio tonal character to fit your musical needs. Whether using the EQ section of a console channel, or an expensive outboard unit, the following material provides a foundation for creating powerful and well-balanced music.

The equalizer or EQ section on a console channel is usually located at about the center of each channel and is definitely one of the most important sections of the mixer. EQ is also called tone control; highs and lows; or highs, mids and lows. Onboard EQ typically has an in/out or bypass button. With the button set to in, your signal goes through the EQ. With the button set to out or bypass, the EQ circuitry is not in the signal path. If you're not using EQ, it is best to bypass the circuit rather than just set all of the controls to flat (no boost and no cut). Any time you bypass a circuit, you eliminate one more possibility for coloration or distortion.

From a purist's standpoint, EQ is to be used sparingly, if at all. Before you use EQ, use the best mic choice and technique. Be sure the instrument you're miking sounds its best. Trying to mike a poorly tuned drum can be a nightmare. It's a fact that you can get wonderful sounds with just the right mic in just the right place on just the right instrument. That's the ideal.

From a practical standpoint, there are many situations where using EQ is the only way to a great sound on time and on budget. This is especially true if you don't own all of the right choice mics. During mixdown, proper use of EQ can be paramount to a really outstanding sound.

Proper control of each instrument's unique tone (also called its timbre) is one of the most musical uses of the mixer, so let's look more closely at equalization.

There are several different types of EQ on the hundreds of different mixers available. What we want to look at are some basic principles that are common to all kinds of boards.

We use EQ for two different purposes:

1. to get rid of (cut) part of the tone that we don't want.

2. to enhance (boost) some part of the tone that we do want.

Hertz

When we use the term Hertz or frequency we're talking about a musical waveform and the number of times it completes its cycle, with a crest and a trough, per second.

Amplitude, Frequency, Length and Speed

Four important characteristics of sound waves are: amplitude, frequency, length and speed.

Amplitude, Frequency, Length and Speed

1. *Amplitude is indicated by the height of the crest and the depth of the trough. Wave B has twice the amplitude of Wave A, moving twice the air.*

2. *Frequency is the number of times the wave completes its 360° cycle in one second. This is expressed as hertz, abbreviated Hz.*

3. *Wavelength can be mathematically calculated as the speed of sound (approximately 1130 feet per second) divided by the frequency of the waveform in Hz (i.e., wavelength=1130÷Hz).*

- *28Hz = 40.36 feet*
  (lowest note on the piano)
- *100Hz = 11.3 feet*
- *1kHz = 1.13 feet*
- *4186Hz = .27 feet or 3.24 inches*
  (highest note on the piano)
- *10kHz = .113 feet or 1.36 inches*
- *20kHz = .0565 feet or .678 inches*

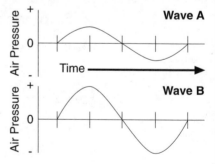

Amplitude expresses the amount of energy in a waveform (the amount of air being moved). In our graphic representation of the sine wave, the amplitude is indicated by the height of the crest and the depth of the trough. Wave B has twice the amplitude of Wave A and, therefore, moves twice the amount of air.

Frequency is the number of times the wave completes its 360° cycle in one second. This is expressed as hertz, abbreviated Hz. One thousand Hz is expressed in kilohertz and is abbreviated kHz. A frequency like 12,000Hz is abbreviated 12kHz or is simply expressed as 12k.

Each frequency has its own length (the number of feet sound travels while it completes one 360° cycle). Wavelength can be mathematically calculated as the speed of sound (approximately 1130 feet per second) divided by the frequency of the waveform in Hz (i.e., wavelength = 1130÷Hz).

As a reference, the human ear can hear a range of frequencies from about 20Hz at the low end to about 20,000Hz at the high end. 20,000Hz is also called 20 kilohertz or 20kHz.

The ability to hear the effect of isolating these frequencies provides a point of reference from which to work. Try to learn the sound of each frequency and the number of hertz that goes with that sound while visualizing a curve with its center point at that frequency.

Listen to the effect that cutting and boosting certain frequencies has on Audio Examples 5 through 13.

Audio Example 5

60Hz

Audio Example 5 starts out flat (meaning no frequencies are cut or boosted). Notice a boost at 60Hz followed by a cut at 60Hz.

Sound Advice on Equalizers, Reverbs & Delays

Equalization Curve

Boosting or cutting a particular frequency also boosts or cuts the frequencies nearby. If you boost 500Hz on an equalizer, 500Hz is the center point of a curve being boosted. Keep in mind that a substantial range of frequencies might be boosted along with the center point of the curve. The exact range of frequencies boosted is dependent upon the shape of the curve.

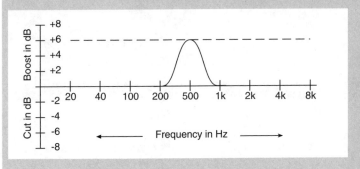

Audio Example 6

120Hz

Audio Example 6 demonstrates a boost then a cut at 120Hz.

Audio Example 7

240Hz

Audio Example 7 demonstrates a boost then a cut at 240Hz.

Audio Example 8

500Hz

Audio Example 8 demonstrates a boost then a cut at 500Hz.

Audio Example 9

1kHz

Audio Example 9 demonstrates a boost then a cut at 1kHz.

Audio Example 10

2kHz

Audio Example 10 demonstrates a boost then a cut at 2kHz.

Audio Example 11

4kHz

Audio Example 11 demonstrates a boost then a cut at 4kHz.

Audio Example 12

8kHz

Audio Example 12 demonstrates a boost then a cut at 8kHz.

Audio Example 13 demonstrates a boost then a cut at 16kHz.

Our goal in understanding and recognizing these frequencies is to be able to create sound pieces that fit together. The frequencies in Audio Examples 5 through 13 represent most of the center points for the sliders on a 10-band graphic EQ.

If the guitar track has many different frequencies, it might sound great all by itself. If the bass track has many different frequencies, it might sound great all by itself. If the keyboard track has many different frequencies, it might sound great all by itself. However, when you put these instruments together in a song, they can get in each other's way and cause problems for the overall mix.

Ideally, we'll find the frequencies that are unnecessary on each track and cut those and then locate frequencies that we like on each track and enhance, or boost, those. If we keep the big picture in mind while selecting frequencies to cut or boost, we can use different frequencies on the different instruments and fit the pieces together.

For instance, if the bass sounds muddy and needs to be cleaned up by cutting at about 250Hz and if the high end of the bass could use a little attack at about 2500Hz, that's great. When we EQ the electric guitar track, it's very possible that we could end up boosting the 250Hz range to add punch. That works great because we've just filled the hole that we created in the bass EQ.

Audio Example 14

Bass (Flat)

Audio Example 14 demonstrates a bass recorded without EQ (flat).

Sound Advice on Equalizers, Reverbs & Delays

Listen to Audio Example 15 as I turn down a frequency with its center point at 250Hz. It sounds much better because I've turned down the frequency range that typically clouds the sound.

Audio Example 15
Bass (Cut 250Hz)

Audio Example 16
Guitar (Flat)

Audio Example 16 demonstrates a guitar recorded flat.

Audio Example 17
Guitar (Boost 250Hz)

Audio Example 17 demonstrates the guitar with a boost at 250Hz. This frequency is typical for adding punch to the guitar sound.

Audio Example 18
Guitar and Bass Together

Audio Example 18 demonstrates the guitar and bass blending together. Notice how each part becomes more understandable as the EQ is inserted.

In a mix, the lead or rhythm guitar doesn't generally need the lower frequencies below about 80Hz. You can cut those frequencies substantially (if not completely), minimizing interference of the guitar's low end with the bass guitar.

If the guitar needs a little grind (edge, presence, etc.) in the high end, select from the 2 to 4kHz range. Since you have already boosted 2.5kHz on the bass guitar, the best choice is to boost 3.5 to 4kHz on guitar. If these frequencies don't work well on the guitar, try shifting the bass high-end EQ slightly. Find different frequencies to boost on each instrument—frequencies that work well together and still sound good on the individual tracks. If you avoid equalizing each instrument at the same frequency, your song will sound smoother and it'll be easier to listen to on more systems.

Definition of Frequency Ranges

The range of frequencies that the human ear can hear is roughly from 20Hz to 20kHz. Individual response may vary, depending on age, climate and how many rock bands the ears' owner might have heard or played in. This broad frequency range is broken down into specific groups. It's necessary for us to know and recognize these ranges.

Listen to Audio Examples 19 through 29 as I isolate these specific ranges.

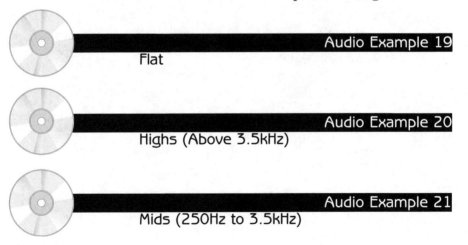

Audio Example 19
Flat

Audio Example 20
Highs (Above 3.5kHz)

Audio Example 21
Mids (250Hz to 3.5kHz)

Audio Example 22

Lows (Below 250Hz)

These are often broken down into more specific categories. Listen to each of these more specific ranges.

Audio Example 23

Flat (Reference)

Audio Example 24

Brilliance

Audio Example 25

Presence

Audio Example 26

Upper Midrange

Audio Example 27

Lower Midrange

Audio Example 28

Bass

Sound Advice on Equalizers, Reverbs & Delays

Frequency Ranges

The range of frequencies that the human ear can hear is roughly from 20Hz to 20kHz. This broad frequency range is broken down into specific groups. It's necessary to know and recognize these ranges.

- **highs**—*above 3.5kHz*
- **mids**—*between 250Hz and 3.5kHz*
- **lows**—*below 250Hz*

These are often broken into more specific categories:

- **brilliance**—*above 6kHz*
- **presence**—*3.5–6kHz*
- **upper midrange**—*1.5–3.5kHz*
- **lower midrange**—*250Hz–1.5kHz*
- **bass**—*60–250Hz*
- **sub-bass**—*below 60Hz*

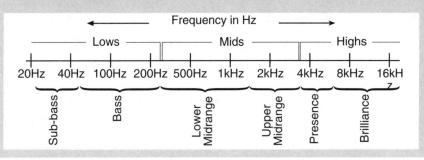

Some of these ranges may be more or less audible on your system, though they're recorded at the same level. Even on the best system, these won't sound equally loud because of the uneven frequency response of the human ear.

Bandwidth

Bandwidth has to do with pinpointing how much of the frequency spectrum is being adjusted. A parametric equalizer has a bandwidth control. Most equalizers that don't have a bandwidth control cut or boost a curve that's about one octave wide. A one-octave bandwidth is specific enough to enable us to get the job done but not so specific that we might create more problems than we eliminate. Bandwidth is sometimes referred to as the Q.

A wide bandwidth (two or more octaves) is good for overall tone coloring. A narrow bandwidth (less than half an octave) is good for finding a problem frequency and cutting it.

Bandwidth (The Q)

Many equalizers provide control over the width of the curve being manipulated. Notice the differing bandwidths in this illustration. Refer to bandwidth in octaves or fractions of an octave.

- *Band #1 is about one octave wide.*
- *Band #2 is about two octaves wide.*
- *Band #3 is about half an octave wide.*

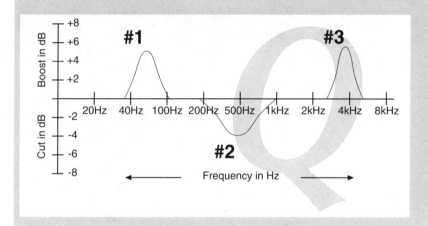

2-Band Onboard EQ

A simple, personal multitrack might only have adjustment for highs and lows. These are each centered on one frequency. Highs are usually between 8 and 10kHz, lows around 100Hz.

This type of EQ can be of some help. For instance, it can help on the guitar track to cut the lows in order to stay out of the way of the bass guitar. Also, on the cymbals you can boost the highs for added brilliance.

If you only have two bands of EQ, consider buying a good outboard graphic or parametric EQ. This isn't as flexible as having great EQ built in to all your mixer channels, but with pre-planning of your arrangement and instrumentation, it is possible to get lots of mileage out of one good EQ. In the 4-track world, pre-planning is the answer to nearly all problems.

Some EQs cut or boost just one fixed frequency. A button near the cut/boost knob can select between two predetermined frequencies.

Selectable Frequencies

Two bands of EQ are available on each knob, enabling access to eight frequency bands. Pressing the Frequency Select button determines which frequency is boosted or cut. Each knob adjusts one frequency or the other, not both at the same time.

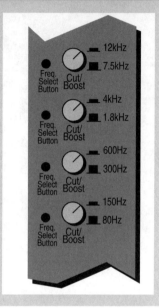

Sweepable EQ

A lot of mixers have sweepable EQ (also called semi-parametric EQ). Sweepable EQ dramatically increases the flexibility of sound shaping. There are two controls per sweepable band:

1. A cut/boost control to turn the selected frequency up or down.

2. A frequency selector that lets you sweep a certain range of frequencies.

This is a very convenient and flexible EQ. With the frequency selector, you can zero in on the exact frequency you need to cut or boost. Often, the kick drum has one sweet spot where the lows are warm and rich or the attack on the guitar is at a very specific frequency. With sweepable EQ, you can set up a boost or cut, then dial in the frequency that breathes life into your music.

Mixers that have sweepable EQ almost always have three separate bands on each channel: one for highs, one for mids and one for lows. Sometimes the highs and lows are fixed-frequency equalizers but the mids are sweepable.

Parametric EQ
This is the most flexible type of EQ. It operates just like a sweepable EQ but provides one other control: the band-width, or Q.

Sweepable EQ

A sweepable equalizer has a cut/boost control to determine the severity of EQ. The frequency selector lets you slide the band throughout a specific range. Equalization like this provides easy and rapid location of just the right frequency band to cut or boost.

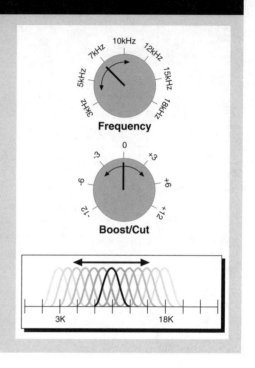

With the bandwidth control, you choose whether you're cutting or boosting a large range of frequencies or a very specific range of frequencies. For example, you might boost a four-octave band centered at 1000Hz, or you might cut a very narrow band of frequencies, a quarter of an octave wide, centered at 1000Hz.

Parametric EQ

The width of the selected frequency band is controlled by the Q adjustment (also called bandwidth). Curve A is a very broad tone control. Curve B is a very specific pinpoint boost. The Q can vary infinitely from its widest bandwidth to its narrowest. Frequency and Boost/Cut operate like the sweepable EQ. The only difference between sweepable and parametric EQ is the addition of the parametric's bandwidth control.

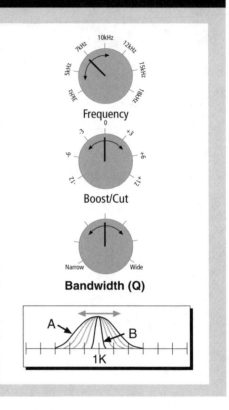

With a tool like this, you can create sonic pieces that fit together like a glove. A parametric equalizer is a great addition to your home studio. They are readily available in outboard configurations and some of the more expensive consoles even have built-in parametric equalization.

These EQs work well together, because where the kick drum has a boost, the bass guitar has a cut, and vice versa. This use of EQ results in the two sounds being independently distinguishable, and it minimizes the risk of accumulating an abundance of one particular frequency.

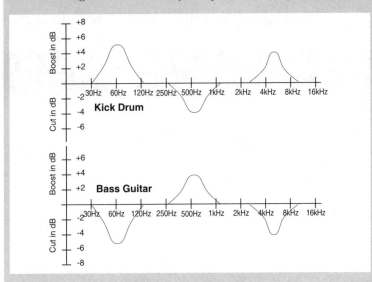

Graphic EQ

This is called a graphic equalizer because it's the most visually graphic of all EQs. It's obvious, at a glance, which frequencies you've boosted or cut.

A graphic equalizer isn't appropriate to include in the channels of a mixer, but it is a standard type of outboard EQ. The graphic EQs that we use in recording have 10, 31 or sometimes 15 individual sliders that each cut or boost a set frequency with a set bandwidth. The bandwidth on a 10-band graphic is one octave. The bandwidth

Graphic Equalizer

The 10-band graphic EQ provides good general sonic shaping. Each slider controls a one octave bandwidth. For more detailed and specific control, try using a 31-band graphic equalizer, where each slider controls one third of an octave.

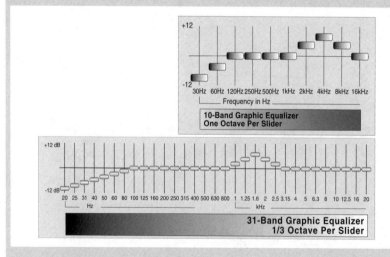

on a 31-band graphic is one third of an octave.

Notch Filter

A notch filter is used to seek and destroy problem frequencies, like a high-end squeal, ground hum or possibly a noise from a heater, fan or camera.

Notch filters have a very narrow bandwidth and are often sweepable. These filters generally cut only.

Highpass Filter

A highpass filter lets the high frequencies pass through unaffected but cuts the low frequencies, usually below about 80Hz. A highpass filter can help minimize 60-cycle hum on a particular track by filtering, or turning down, the fundamental frequency of the hum.

Highpass filters function very well when you need to eliminate an ambient rumble, like a furnace in the background or street noise that leaks into a vocal mic.

Lowpass Filter

A lowpass filter lets the low frequencies pass through unaffected and cuts the highs, usually above about 8 to 10kHz. These filters have many uses. For instance, a lowpass filter can help minimize cymbal leakage onto the tom tracks, filter out a high buzz in a guitar amp or filter out string noise on a bass guitar track.

Highpass and lowpass filters are a specific type of equalizer called a shelving equalizer. A shelving EQ leaves all frequencies flat to a certain point, then turns all frequencies above or below that point down or up at a rate specified in dB per octave. Most high- and lowpass filters roll off the highs or lows at a rate between 6 and 12dB per octave.

Shelving equalizers are useful when trying to get rid of unnecessary frequencies on the individual tracks. For example, on a bass guitar or kick drum track, the frequencies above 8k or so are typically useless. Applying the lowpass filter here could help get rid of any tape noise or leakage in the high frequencies on these tracks.

Use of these filters might be barely noticeable. That's good because it indicates that we're keeping the part of the signal we want and filtering out the frequencies that we don't need. Some mixers include sweepable shelving EQ (adjustable high- and lowpass filters). With sweepable shelving EQ, you can carve away at the highs and lows of each track to totally eliminate unnecessary frequency conflicts. Be sure, though, that you're not robbing your music of life-giving upper harmonics.

Remember, even with the multitude of equalizers available, don't use EQ first to shape your sounds. First, get as close to the sound you want using mic choice and mic technique, then use EQ if it's necessary.

Delay Effects

A delay does just what its name says: it hears a sound and then waits for a while before it reproduces it. Current delays are simply digital recorders that digitally record the incoming signal and then play it back with a time delay selected by the user. This time delay can vary from unit to unit, but most delays have a range of delay length from a portion of a millisecond up to one or more seconds. This is called the delay time or delay length and is variable in milliseconds.

Almost all digital delays are much more than simple echo units. Within the delay are all of the controls you need to produce slapback, repeating echo, doubling, chorusing, flanging, phase shifting, some primitive reverb sounds and any hybrid variation you can dream up.

Slapback Delay

The simplest form of delay is called a slapback. The slapback delay is a single repeat of the signal. Its delay time is anything above about 35ms. Any single repeat with a delay time of less than 35ms is called a double.

To achieve a slapback from a delay, simply adjust the delay time and turn the delayed signal up, either on the return channel or on the mix control within the delay.

For a single slapback delay, feedback and modulation are set to their off positions. Slapback delays of between 150ms and about 300ms are very effective and common for creating a big vocal or guitar sound.

Audio Example 30

250ms Slapback

Audio Example 30 demonstrates a track with a 250ms slapback delay.

Slapback delays between 35 and 75ms are very effective for thickening a vocal or instrumental sound.

Audio Example 31

50ms Slapback

Audio Example 31 demonstrates a track with a 50ms delay.

Slapback delay can be turned into a repeating delay. This smooths out the sound of a track even more and is accomplished through the use of the regeneration control. This is also called feedback or repeat.

This control takes the delayed signal and feeds it back into the input of the delay unit, so we hear the original, the delay and then a delay of the delayed signal. The higher you turn the feedback up, the more times the delay is repeated. Practically speaking, anything past about three repeats gets too muddy and does more musical harm than good.

Sound Advice on Equalizers, Reverbs & Delays

The vocal track in Audio Example 32 starts with a simple single slapback, then the feedback raises until we hear three or four repeats.

Why does a simple delay make a track sound so much bigger and better? Delay gives the brain the perception of listening in a larger, more interesting environment. As the delays combine with the original sound, the harmonics of each part combine in interesting ways. Any pitch discrepancies are averaged out as the delay combines with the original signal. If a note was sharp or flat, it's hidden when heard along with the delay of a previous note that was in tune. This helps most vocal sounds tremendously and adds to the richness and fullness of the mix.

The human brain gets its cue for room size from the initial reflections, or repeats, that it hears off surrounding surfaces. Longer delay times indicate, to the brain, that the room is larger. The slapback is really perceived as the reflection off the back wall of the room or auditorium as the sound bounces back (slaps back) to the performer. Many great lead vocal tracks have used a simple slapback delay as the primary or only effect. Frequently, this delay sounds cleaner than reverb and has less of a tendency to intrusively accumulate.

Slapback delay is typically related in some way to the beat and tempo of the song. The delay is often in time with the eighth note or sixteenth note, but it's also common to hear a slapback in time with the quarter note or some triplet subdivision. The delay time can add to the rhythmic feel of the song. A delay that's in time with the eighth note can really smooth out the groove of the song, or if the delay time is shortened or lengthened just slightly, the groove may feel more aggressive or relaxed. Experiment with slight changes in delay time.

It's easy to calculate the delay time, in milliseconds, for the quarter note in your song, especially when you're working from a sequence and the tempo is already available on screen. Simply divide 60,000 by the tempo of your song (in beats per minute). 60,000 ÷ bpm = delay time per beat in milliseconds (typically the quarter note).

Doubling

A single delay of less than 35ms is called a double. This short delay can combine with the original track to sound like two people (or instruments) on the same part. Often, performers will actually record the same part twice to achieve the doubled sound, but sometimes the electronic double is quicker, easier and sounds more precise. Audio Example 33 demonstrates an 11ms delay (with no feedback and no modulation) combined with the original vocal. At the end of the example, the original and the delayed double pan apart in the stereo spectrum. This can be a great sound in stereo, but is a potential problem when summing to mono.

Audio Example 33

11ms Vocal Delay

When doubling, use prime numbers for delay times. You'll hear better results when your song is played in mono. A prime number can only be divided by one and itself (e.g., 1, 3, 5, 7, 11, 13, 17, 19, 23, 29 and so on).

Modulation

The modulation control on a delay is for creating chorusing, flanging and phase shifting effects. The key factor here is the LFO (low frequency oscillator); its function is to continually vary the delay time. The LFO is usually capable of varying the delay from the setting indicated by the delay time to half of that value and back.

Sound Advice on Equalizers, Reverbs & Delays

Sometimes the LFO control is labeled modulation.

As the LFO is slowing down and speeding up the delay, it's speeding up and slowing down the playback of the delayed signal. In other words, modulation actually lowers and raises the pitch in exactly the same way that a tape recorder does if the speed is lowered and raised. Audio Example 34 demonstrates the sound of the LFO varying the delay time. This example starts subtly, with the variation from the original going down slightly, then back up. Finally, the LFO varies dramatically downward, then back up again.

Audio Example 34

The LFO

On most usable effects, these changes in pitch are slight and still within the boundaries of acceptable intonation, so they aren't making the instrument sound out of tune. In fact, the slight pitch change can have the effect of smoothing out any pitch problems on a track.

As the pitch is raised and lowered, the sound waves are shortened and lengthened. When two waveforms follow the same path, they sum together. The result is twice the amount of energy. In addition, when two waveforms are out of phase, they work against and cancel each other, either totally or partially.

When the modulation is lengthening and shortening the waveform and the resulting sound is combined with the original signal, the two waveforms continually react together in a changing phase relationship. They sum and cancel at varying frequencies. The interaction between the original sound and the modulated delay can simulate the sound we hear when several different instrumentalists or vocalists perform together. Even though each member of a choir tries their hardest to stay in tune and together rhythmically, they're continually varying pitch and timing. These variations are like the interaction of the modulated delay with the original track. The chorus setting on an effects processor is simulating the sound of a real choir by combining the original signal with the modulated signal.

The speed control adjusts how fast the pitch raises and lowers. These changes might happen very slowly, taking a few seconds to complete one cycle of raising and lower the pitch, or they might happen quickly, raising and lowering the pitch several times per second.

Audio Example 35 demonstrates the extreme settings of speed and depth. It's obvious when the speed and depth controls are changed here. Sounds like these aren't normally used, but when we're using a chorus, flanger or phase shifter, this is exactly what is happening, in moderation.

Extreme Speed and Depth

Phase Shifter

Now that we're seeing what all these controls do, it's time to use them all together. Obviously, the delay time is the key player in determining the way that the depth and speed react. If the delay time is very, very short, in the neighborhood of 1ms or so, the depth control will produce no pitch change. When the original and affected sounds are combined, we hear a distinct sweep that sounds more like an EQ frequency sweeping the mids and highs. With these short delay times, we're really simulating waveforms, moving in and out of phase, unlike the larger changes of singers varying in pitch and timing. The phase shifter is the most subtle, sweeping effect, and it often produces a swooshing sound.

Phase Shifter

Audio Example 36 demonstrates the sound of a phase shifter.

Flanger

A flanger has a sound similar to the phase shifter, except it has more variation and color. The primary delay setting on a flanger is typically about 20ms. The LFO varies the delay from near 0ms to 20ms and back, continually. Adjust the speed to your own taste.

Flangers and phase shifters work very well on guitars and Rhodes-type keyboard sounds.

Flanger

Audio Example 37 demonstrates the sound of a flanger.

Chorus

The factor that differentiates a chorus from the other delay effects is, again, the delay time. The typical delay time for a chorus is about 15 to 35ms, with the LFO and speed set for the richest effect for the particular instrument voice or song. With these longer delay times, as the LFO varies we actually hear a slight pitch change. The longer delays also create more of a difference in attack time. This also enhances the chorus effect. Since the chorus gets its name from the fact that it's simulating the pitch and time variation that exist within a choir, it might seem obvious that a chorus works great on background vocals. It does. Chorus is also an excellent effect for guitar and keyboard sounds.

Audio Example 38

Chorus

Audio Example 38 demonstrates the sound of a chorus.

Phase Reversal and Regeneration

The regeneration control can give us multiple repeats by feeding the delay back into the input so that it can be delayed again. This control can also be used on the phase shifter, chorus and flange. Regeneration, also called feedback, can make the effect more extreme or give the music a sci-fi feel. As you practice creating these effects with your equipment, experiment with feedback to find your own sounds.

Most units have a phase reversal switch that inverts the phase of the affected signal. Inverting the phase of the delay can cause very extreme effects when combined with the original signal (especially on phase shifter and flanger effects). This can make your music sound like it's turning inside out.

Audio Example 39 begins with the flanger in phase. Notice what happens to the sound as the phase of the effect is inverted.

Audio Example 39
Inverting Phase

Stereo Effects

The majority of effects processors are stereo, and with a stereo unit, different delay times can be assigned to the left and the right sides. If you are creating a stereo chorus, simply set one side to a delay time between 15 and 35ms, then set the other side to a different delay time, between 15 and 35ms. All of the rest of the controls are adjusted in the same way as a mono chorus. The returns from the processor can then be panned apart in the mix for a very wide and extreme effect. Listen as the chorus in Audio Example 40 pans from mono to stereo.

Audio Example 40
Stereo Chorus

For a stereo phase and flange, use the same procedure. Simply select different delay times for the left and right sides.

Understanding what is happening within a delay is important when you're trying to shape sounds for your music.

Sometimes it's easiest to bake a cake by simply pressing the Bake Me a Cake button, but if you are really trying to create a meal that flows together perfectly, you might need to adjust the recipe for the cake. That's what we need to do when building a song, mix or arrangement; we must be able to custom fit the pieces.

Reverberation Effects

As we move from the delay effects into the reverb effects, we must first realize that reverb is just a series of delays. Reverberation is simulation of sound in an acoustical environment, like a concert hall, gymnasium or bedroom.

No two rooms sound exactly alike. Sound bounces back from all the surfaces in a room to the listener or the microphone. These bounces are called reflections. The combination of the direct and reflected sound in a room creates a distinct tonal character for each acoustical environment. Each one of the reflections in a room is like a single delay from a digital delay. When it bounces around the room, we get the effect of regeneration. When we take a single short delay and regenerate it many times, we're creating the basics of reverberation. Audio Example 41 demonstrates the unappealing sound of simulated reverb, using a single delay.

Audio Example 41
Simulated Reverb

Reverb must have many delays and regenerations combining at once to create a smooth and appealing room sound. Audio Example 42 demonstrates the smooth quality created by many delays working together in the proper balance.

Reverberation

If you can envision thousands of delays bouncing (reflecting) off thousands of surfaces in a room and then back to you, the listener, that's what's happening in the reverberation of a concert hall or any acoustical environment. There are so many reflections happening in such a complex order that we can no longer distinguish individual echoes.

Our digital simulation of this process is accomplished by a digital reverb that produces enough delays and echoes to imitate the smooth sound of natural reverb in a room. The reason different reverb settings sound unique is because of the different combinations of delays and regenerations.

A digital reverb is capable of imitating a lot of different acoustical environments and can do so with amazing clarity and accuracy. The many different echoes and repeats produce a rich and full sound. Digital reverbs can also shape many special effects that would never occur acoustically. In fact, these sounds can be so fun to listen to that it's hard not to overuse reverb.

Keep in mind that sound perception is not just two dimensional, left and right. Sound perception is at least three dimensional, with the third dimension being depth (distance). Depth is created by the use of delays and reverb. If a sound (or a mix) has too much reverb, it loses the feeling of closeness or intimacy and sounds like it's at the far end of a gymnasium. Use enough effect to achieve the desired results, but don't overuse effects.

Most digital reverbs have several different reverb sounds available. These are usually labeled with descriptive names like halls, plates, chambers, rooms, etc.

Slapback Delay and Reflections

Sound travels at the rate of about 1130 ft./sec. To calculate the amount of time (in seconds) it takes for sound to travel a specific distance, divide the distance (in feet) by 1130 (ft./sec.): time = distance (ft.) ÷ speed (1130 ft./sec.). In a 100' long room, sound takes about 88ms to get from one end to the other (100÷1130).

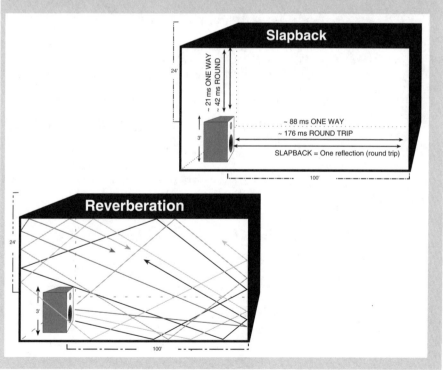

Hall Reverb

Hall indicates a concert hall sound. These are the smoothest and richest of the reverb settings. Audio Example 43 demonstrates a Hall Reverb.

Audio Example 43
Hall Reverb

Chamber Reverb

Chambers imitate the sound of an acoustical reverberation chamber, sometimes called an echo chamber. Acoustical chambers are fairly large rooms with hard surfaces. Music is played into the room through high-quality, large speakers, and then a microphone in the chamber is patched into a channel of the mixer as an effects return. Chambers aren't very common now that technology is giving us great sounds without taking up so much real estate. The sound of a chamber is smooth, like the hall's, but has a few more mids and highs.

Audio Example 44 demonstrates the sound of a chamber reverb. Compare this sound to the concert hall reverb in Audio Example 43.

Audio Example 44
Chamber Reverb

Plate Reverb

Plates are the brightest sounding of the reverbs. These sounds imitate a true plate reverb. A true plate is a large sheet of metal (about 4' by 8') suspended in a box and allowed to vibrate freely. Sound is induced onto the plate by a speaker attached to the plate itself. Two contact microphones are typically mounted on the plate at different locations to give a stereo return to the mixer. The sound of a true plate reverb has lots of highs and is very clean and nonintrusive.

A digital simulation of the plate is also full of clean highs. Audio Example 45 demonstrates a plate reverb sound.

Plate Reverb

Room Reverb

A room setting can imitate many different types of rooms that are typically smaller than the hall/chamber sounds. These can range from a bedroom to a large conference room or a small bathroom with lots of towels to a large bathroom with lots of tile.

Rooms with lots of soft surfaces have little high-frequency content in their reverberation. Rooms with lots of hard surfaces have lots of high-frequency in their reverberation.

Audio Example 46 demonstrates some different room sounds. Review and compare the previous examples of reverberation types.

Audio Example 46

Room Reverb

Reverse Reverb

Most modern reverbs include reverse or inverse reverb. These are simply backwards reverb. After the original sound is heard, the reverb swells and stops. It is turned around. These can actually be fairly effective and useful if used in the appropriate context.

Audio Example 47 demonstrates reverse reverb.

Audio Example 47
Reverse Reverb

Gated Reverb

Gated reverbs have a sound that is very intense for a period of time, then closes off quickly. This has become very popular because it can give an instrument a very big sound without overwhelming the mix, because the gate keeps closing it off.

Though this technique comes and goes with trends and styles it has been around for a long time. Audio Example 48 demonstrates a gated reverb sound.

There are many variations of labels for reverb. You might see bright halls, rich plates, dark plates, large rooms, small rooms, or bright phone booth, but they can all be traced back to the basic sounds of halls, chambers, plates and rooms.

These sounds often have adjustable parameters. They let us shape the sounds to our music so that we can use the technology as completely as possible to enhance the artistic vision. We need to consider these variables so that we can customize and shape the effects.

Predelay

Predelay is a time delay that happens before the reverb is heard. This can be a substantial time delay (up to a second or two) or just a few milliseconds. The track is heard clean (dry) first, so the listener can get a more upfront feel; the reverb comes along shortly thereafter to fill in the holes and add richness. Listen to Audio Example 49 as the predelay setting is changed.

Audio Example 49
Predelay

Diffusion

Diffusion controls the space between the reflections.

A low diffusion can be equated with a very grainy photograph. We might even hear individual repeats in the reverb.

A high diffusion can be equated with a very fine grain photograph, and the sound is a very smooth wash of reverb.

Listen to the reverberation in Audio Example 50 as the change is made from low diffusion to high diffusion.

Diffusion

Decay Time

Reverberation time, reverb time and decay time all refer to the same thing. Traditionally, reverberation time is defined as the time it takes for the sound to decrease to one-millionth of its original sound pressure level. In other words, it's the time it takes for the reverb to go away.

Decay time can typically be adjusted from about 1/10 of a second up to about 99 seconds. We have ample control over the reverberation time. Audio Example 51 demonstrates a constant reverb sound with a changing decay time.

Decay Time

Density

The density control adjusts the initial short delay times. Low density is good for smooth sounds like strings or organ. High density works best on percussive sounds.

Diffusion

Reverberation with low diffusion often sounds grainy because the reflections that make up the reverb sound are relatively far apart. High diffusion usually results in a smooth-sounding reverb. Since the reflections are close together in time, they blend together to create a wash of reverberation.

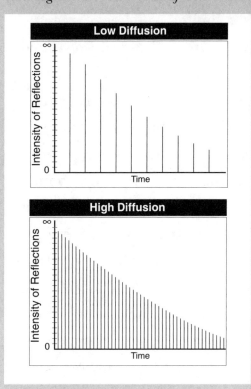

Should I Print Reverb or Delay to Tape?

It is ideal to save additions of reverb and delay until mixdown. It's common to set up an effects bus for monitor purposes only during tracking that isn't recorded. This helps the artist feel better about the emotion of the performance.

If you have a small 4-track or 8-track setup and a limited number of effects, you'll often be forced to print reverb and delay along with the instrument or voice on the same track. This is a commitment. Select the blend of the wet and dry signals that sounds good at the time you record the part. If you have a question about the amount of reverb or delay to record, use the least amount that you think you need. If you print too much delay or reverb, it's there forever. Too much delay and reverb can make a part sound like it was recorded from the far end of a gymnasium, especially as the mix develops. The only real way to fix this is to rerecord the part with less effect, which can be costly and is, at the very least, a nuisance. If you record the part a little on the dry side, you can always add whatever reverb is readily available during mixdown to make it sound more distant.

Density

Low-density reverberation is good for smooth sounds like a string pad or a mellow organ. The increased spacing of the initial reflections can sound grainy on sounds with percussive attacks. High-density reverberation sounds good on percussive instruments. The closer spacing of the initial reflections produces a smooth-sounding reverb on percussive attacks.

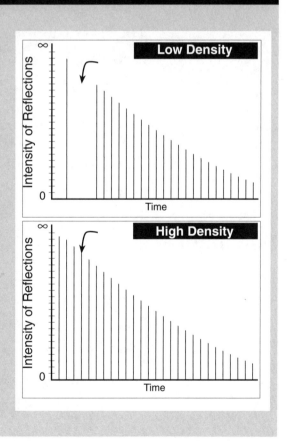

Be flexible. If a player has come up with a great sound that might take you a while to duplicate, and if they want to print the sound to tape, give it a try, but be conservative. Conscientious musicians often come up with great and interesting variations of a sound, and much of the final musical impact depends on how the signal is patched through the effects. They may run the chorus through the distortion or the distortion through the chorus. Both sound different. These routing changes can really result in some unique sounds. Take advantage of the player's diligence. You'll all share the benefit of a great sound.

Summary

In the recording world of yesteryear, the only way to adjust reverberation time was to physically change absorption in a physical echo chamber (room) or to dampen or undampen the springs in the spring reverb tanks or physically move a bar that

Sound Advice on Equalizers, Reverbs & Delays

moved a felt pad onto, or off of, the plate reverb. Current technology provides a myriad of variables when shaping reverb sounds. In fact, when you consider the number of possible options, it can be mind boggling. We can design unnatural hybrids like a large room with a very short decay time and plenty of high frequency, or any other natural or unnatural effect.

Each equalization, delay or reverb parameter is important. As we deal with individual guitar, drum, keyboard and vocal sounds, we must possess understanding that provides depth and insight to the final goal. If you expect to have professional-sounding recordings, you must be able to use all the tools available to you in a way that supports and enhances the emotional and artistic power of the music. Assess what you're doing and how you're doing it, then do your best to "take it a little further."

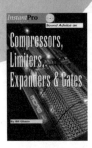

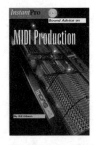